D1368641

PRIVACY IN THE DIGITAL AGE

PERSONAL
DATA COLLECTION

BY A.W. BUCKEY

CONTENT CONSULTANT
M. E. Kabay, PhD, CISSP-ISSMP
Professor of Computer Information Systems
Norwich University

Core Library

An Imprint of Abdo Publishing
abdobooks.com

Cover image: Many companies and organizations collect people's personal data.

abdocorelibrary.com

Published by Abdo Publishing, a division of ABDO, PO Box 398166,
Minneapolis, Minnesota 55439. Copyright © 2020 by Abdo Consulting
Group, Inc. International copyrights reserved in all countries. No part of this
book may be reproduced in any form without written permission from the
publisher. Core Library™ is a trademark and logo of Abdo Publishing.

Printed in the United States of America, North Mankato, Minnesota
032019
092019

Cover Photo: Shutterstock Images
Interior Photos: Shutterstock Images, 1, 20, 32–33, 45; Kaspars Grinvalds/Shutterstock Images,
4–5; Red Line Editorial, 8, 27; Everett Historical/Shutterstock Images, 10–11, 43; Win McNamee/
Getty Images News/Getty Images, 12–13; iStockphoto, 16–17, 24–25; Dake Kang/AP Images,
19; Patrick Semansky/AP Images, 22; Jose Luis Magana/AP Images, 29; Alexandros Michailidis/
Shutterstock Images, 36–37; Ipon-Boness/SIPA/Newscom, 38; Toby Talbot/AP Images, 40

Editor: Maddie Spalding
Series Designer: Megan Ellis

Library of Congress Control Number: 2018966006

Publisher's Cataloging-in-Publication Data

Names: Buckey, A. W., author.
Title: Personal data collection / by A. W. Buckey
Description: Minneapolis, Minnesota: Abdo Publishing, 2020 | Series: Privacy in the digital age |
        Includes online resources and index.
Identifiers: ISBN 9781532118937 (lib. bdg.) | ISBN 9781532173110 (ebook) | ISBN
        9781644940846 (pbk.)
Subjects: LCSH: Personal data protection--Juvenile literature. | Computer systems--Protection
        --Juvenile literature. | World Wide Web--Security measures--Juvenile literature.
        | Privacy, Right of--United States--Juvenile literature. | Personal information
        management-- Juvenile literature.
Classification: DDC 005.8--dc23

CONTENTS

CHAPTER
ONE

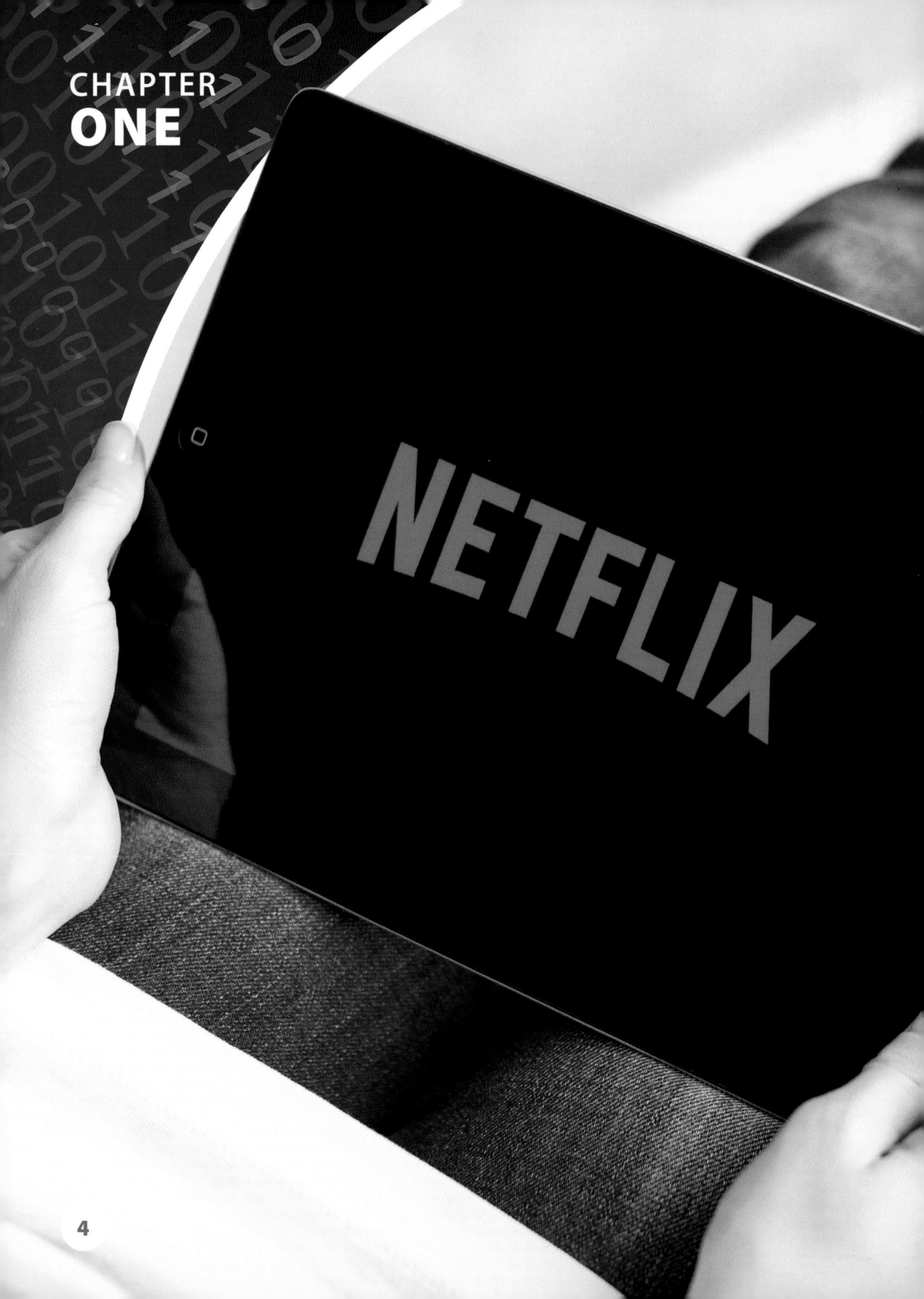

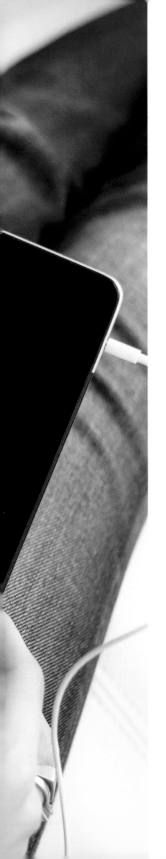

RECOMMENDED FOR YOU

t is a Friday night. Diya wants to watch a movie. She turns on Netflix. She sees rows of movies on the Netflix home screen. Each one comes with a description of the plot. They also have starred ratings. Diya can see if other Netflix users liked the movie she is considering. She has a lot of information at her fingertips. But Netflix is also collecting information on her.

The videos Diya sees were chosen for her. Other users see different options on their home screens. When Diya chooses a movie, Netflix records her choice. If she pauses or stops the movie, Netflix records that too. Diya may see new options on her home screen the

Netflix gathers data from its viewers.

next time she opens Netflix. The company updates recommendations based on the information it gathers.

Netflix collects and stores information on all of its users. Information that can be collected and stored is called data. Netflix compares users' data. It tracks which people like the same shows and movies. Netflix uses this data to put people into groups. It calls these groups "taste clusters." Netflix decides which new videos each cluster might enjoy.

USING DATA

Netflix produces movies and shows. It wants its programs to be popular. The data

NIELSEN RATINGS

A company called Nielsen collects data on television users. It started doing this in 1950. It collects data through special devices. These devices are attached to televisions across the country. The devices gather data on which TV shows are watched most often. Nielsen ratings used to be a popular source of data. They are less popular today than they used to be. Today, TV viewership is changing. Many people watch TV shows through streaming services such as Netflix and Hulu.

Netflix collects helps it make business decisions. Netflix uses data to determine which shows to keep and which shows to cancel. For example, Netflix produces a scary show called *Black Mirror*. This show is a hit with many clusters. Some clusters that do not usually like scary shows watch *Black Mirror*. Netflix can use this data to keep making shows that are likely to be popular. Netflix's data usage strategy has worked well. The company estimates that it has made $1 billion from using personal data.

TYPES OF PERSONAL
DATA

The diagram below shows some types of personal data that companies may collect. Why do you think companies collect a variety of data? Which data would you want to protect most?

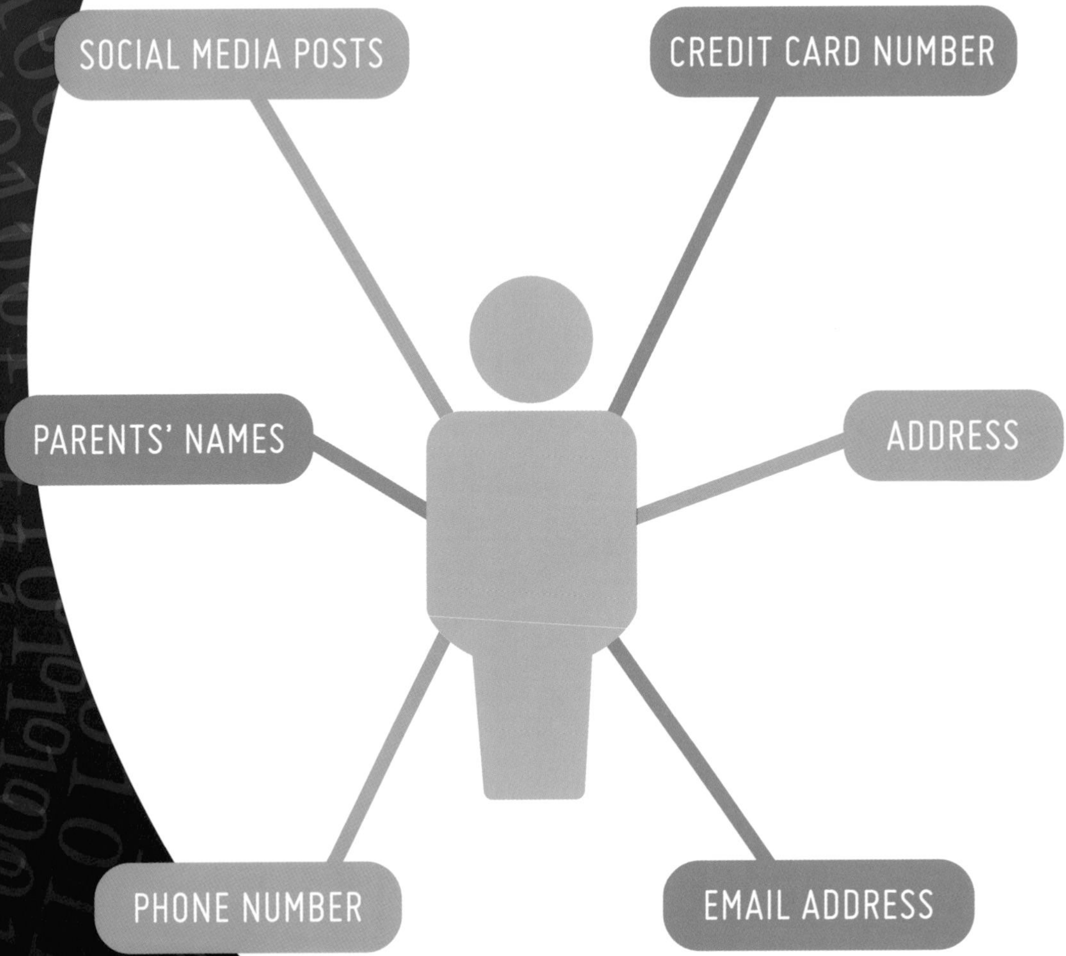

SOCIAL MEDIA POSTS

CREDIT CARD NUMBER

PARENTS' NAMES

ADDRESS

PHONE NUMBER

EMAIL ADDRESS

Personal data is information about a person. It includes a person's likes and dislikes. Other types of personal data include a person's job and medical history. Companies and governments collect personal data. Data collection has many uses. It is a tool for researching people. Companies collect personal data to improve their businesses. The government and the police also use personal data.

Today, collecting personal data is easier than ever. Some people think personal data collection interferes with their right to privacy. Others worry that their data will be used in unfair ways. Data can be used to discriminate against people. Some companies collect personal data and sell it for profit. These practices raise concerns about data privacy.

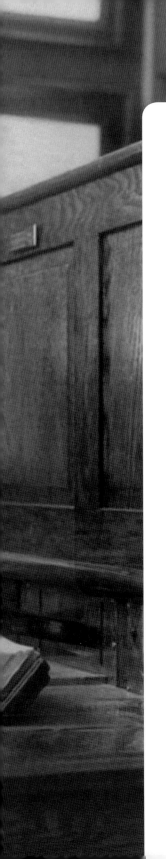

THE HISTORY OF DATA COLLECTION

Data collection has a long history. Some of the first data collection projects were censuses. A census is an official count of a population. Ancient Romans used a census to find out what citizens owned. They used this data to figure out how much to tax people. They taxed rich people more than poor people.

Over time, new data collection projects emerged. English researcher John Graunt studied personal data in the mid-1600s.

In the early 1900s, workers at the US Census Bureau used special machines to process census data.

NEW DATA

It is impossible to analyze all of the new data that is created. People must choose which data to study. Some people see the wealth of data as a tool for progress. For example, data can help doctors find the best treatments for their patients. But people and companies can misuse data. Companies could use data to discriminate against people. Databases can be vulnerable. Hackers can break into databases and steal personal data. This is called a data breach.

This data included age and lifestyle choices. He used this data to estimate life expectancy. Life expectancy is the average number of years a person can expect to live. It is estimated by studying a large population of people.

Today, researchers still use personal data to predict life expectancy. Census data continues to be valuable too. Researchers at the US Census Bureau collect census data. This includes

The US Census Bureau collects and publishes data about the entire US population every ten years.

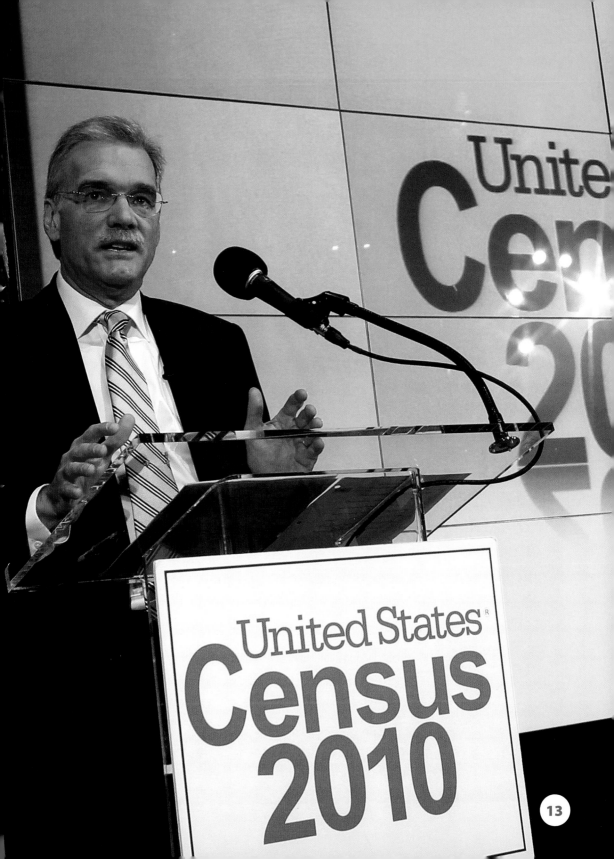

United States^R
Census
2010

data about how many people live in certain areas. It also includes data about different populations, such as their jobs and education. Researchers use this data to find out which areas need new services or facilities, such as schools.

THE DATA EXPLOSION

The term *big data* was invented in the late 1980s. It refers to the data explosion made possible by the internet. Big data is vast and always growing.

Programs that keep track of user

activity create and collect data. Computer systems called databases store data. For example, libraries often have databases. The database shows which books are available. It can be found on the library's website.

The internet has made it easier to create new types of data. Almost every online activity creates personal data. Databases save a person's web searches. They also collect other kinds of data, such as a person's online shopping history.

EXPLORE ONLINE

Chapter Two talks about how data is stored. The article at the website below goes into more depth on this topic. Does the article answer any of the questions you had about data storage?

HOW IS DATA STORED ON A COMPUTER?
abdocorelibrary.com/personal-data-collection

DATA RESEARCH

Most data is not useful on its own. But it can become valuable when people use it for research. Data research reveals patterns. It can help people make predictions. These predictions can help improve people's lives.

HEALTH DATA

Some researchers study health data to make predictions. Health data often comes from medical records. Most US hospitals have Electronic Health Records (EHRs). EHRs are records of patients' data. This data is stored in databases.

Doctors record patient health data in electronic health records.

RESEARCHER BIAS

Researchers may have biases. Bias occurs when people favor one point of view over others. It can affect the type of data researchers collect. It can also affect the way researchers collect data. One common error is sampling bias. This occurs when researchers collect data from a sample that does not represent a population. For example, researchers may only collect data from people in a certain racial group. The larger population may be made up of many different races. The sample does not represent that population. The researchers' conclusions might not apply to everyone within the population.

In 2015 the US government started a project with EHR data. The project is called the *All of Us* research program. Scientists are collecting the EHRs of 1 million volunteers. They are studying this data to find patterns. People living very different lives may still have similar health histories. Scientists hope to find new risk factors for diseases. Then doctors can watch out for these risk factors in patients. This could help doctors make predictions and prevent diseases.

A researcher working on the *All of Us* program collects health data from a volunteer.

OTHER USES

Data research can be used in many ways. For example, the life insurance business depends on health data. Life insurance companies sell policies. Customers buy policies worth large sums of money. They make yearly payments to the company. If a customer dies, the company pays a sum of money to the customer's loved ones. Life insurance companies try to make money from policies. They research health data. They often seek out healthy people as customers. Healthy people tend to live for many years. They make payments on their policies each year. For these reasons, they make good life insurance customers.

Some life insurance companies collect customers' health data. They give discounts to customers who share their health data. Devices such as fitness trackers collect this information. A fitness tracker is usually worn on a person's wrist. It counts the number of steps a person takes each day. It can also track other exercise.

PRIVACY AND FAIRNESS

There are drawbacks to easy data sharing. People may want their personal data to remain private. Health data can be very sensitive.

Some people use fitness trackers to collect health data.

The National Security Agency collects data from phone call records.

Many people do not want to share this data. Companies often make personal data anonymous. They remove names from data to protect people's identities. But even anonymous data can sometimes be traced back to its owners. Some people consent to sharing some of their data. But companies may not tell people all the ways that their data may be used.

Data can also be chosen and collected unfairly. For example, not everyone can use a fitness tracker.

Some people are elderly or in poor health. They may not be able to walk long distances. This may exclude them from fitness discounts.

Some people have concerns about how governments collect data. The National Security Agency (NSA) monitors people who might be security threats. This includes suspected terrorists and people who are linked to suspects. The NSA collects phone call data from millions of Americans. Some people think the NSA monitors more people than is necessary. They believe the NSA violates people's privacy rights.

FURTHER EVIDENCE

Chapter Three explains how data can help with research. What was one of the main points of this chapter? What evidence is included to support this point? The article at the website below explains how and why some governments collect personal data. Does the information on the website support this point? Does it present new evidence?

METHODS AND PROBLEMS OF DATA COLLECTION
abdocorelibrary.com/personal-data-collection

SELLING AND SHARING DATA

Most internet activities create data. A computer downloads cookies when a user visits a website. Most sites use cookies. Cookies are files. They track the sites a person visits. They report this data back to the site they came from. This may be a company's site. The company can use this data. It can sell the data to other companies. It can also sell access to the data. For example, some companies pay the social media site Twitter for access to its user data. Companies study this data to learn how to better market their products.

Websites use cookies to collect data when people visit them.

DATA SALES

Some companies specialize in selling data. These companies are called data brokers. Data brokers often collect consumer data. Consumer data is information about what people buy. Data brokers separate people into groups based on this data. They group together people who have things in common. For example, they may group people of similar ages together. People within a group often buy similar things. Companies can use these groups to figure out what products their customers might want.

Other companies collect and store their own data. For example, Facebook saves every message sent over its site. It tracks posts

MOBILE APPS

Mobile apps are programs for mobile devices. Mobile apps such as Uber and Google Maps provide services. Uber provides a ride-sharing service. Google Maps provides navigation. Other popular apps are games, such as *Fortnite*. A 2015 study found that most mobile apps share user data with other companies.

FACEBOOK AD SALES

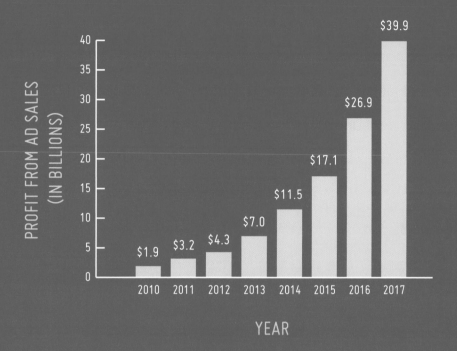

PROFIT FROM AD SALES (IN BILLIONS)

YEAR

Year	Value
2010	$1.9
2011	$3.2
2012	$4.3
2013	$7.0
2014	$11.5
2015	$17.1
2016	$26.9
2017	$39.9

The above graph shows how much money Facebook made from ad sales in certain years. What trend do you notice? How do you think Facebook's data collection is related to its ad sales?

that users like and share. It also tracks users' friend networks. This data gives Facebook information about users' interests. Facebook uses this data to figure out which advertisements a user may like. Companies pay Facebook to show ads. Ads appear on the news feeds of users that are most likely to buy the products. This strategy is called targeted advertising.

RISKS OF SELLING DATA

People can choose to share and sell their data. But they cannot always control how companies will use it. Data can also be used in unexpected ways. For example, Facebook data played a role in the 2016 US presidential election. In 2013 a data scientist created an app that was bought by the company Cambridge Analytica. People used the app to take personality quizzes. The app collected data from millions of Facebook users. It also got access to the users' friends' data.

Donald Trump was a candidate in the 2016 election. Trump's campaign team hired Cambridge Analytica. The company used the data it had collected to predict who might vote for Trump. It sent targeted ads to these people through Facebook. The Cambridge Analytica scandal raised important privacy questions. Facebook has since updated its data sharing rules. Companies can

In April 2018, Facebook chief executive officer Mark Zuckerberg spoke to the US Congress about how the company would address data privacy issues.

still collect Facebook user data. But they can only collect data with users' permission.

Facebook controls which companies have access to its data. But it cannot control how other companies use this data. Many people think this is unfair. They argue that people deserve more information about targeted ads. They want companies to tell them more about how and when their data is collected.

STRAIGHT TO THE
SOURCE

Steven Petrow is a writer and cancer survivor. His medical data was used and sold without his knowledge. In a 2018 article for the *Washington Post*, he wrote:

> *It's a good bet that the fine print of the consent form you signed before your latest test or operation said that all the data or tissue samples belong to the doctor or institution performing it. They can study it, sell it or do whatever they want with it, without notifying or compensating you, although the data must be depersonalized. . . .*
>
> *As a cancer survivor, I know the importance of advancing medical science, fast. . . . But I had paid only cursory attention to the consent form I was given before treatment, mostly because I was crazy-scared.*
>
> Source: Steven Petrow. "Who Owns Your Medical Data? Most Likely Not You." *The Washington Post*. The Washington Post, November 25, 2018. Web. Accessed December 19, 2018.

Consider Your Audience
Adapt this passage for a different audience, such as your friends. Write a blog post conveying this same information for the new audience. How does your post differ from the original text and why?

Welcome to yo

saliva collectio

THE FUTURE OF DATA COLLECTION

The data explosion is not slowing down. People are creating new types of personal data. Some people think this data explosion could jeopardize many more people's information. Lawmakers are creating personal data laws. These laws aim to protect people's privacy and guard against the misuse of personal data.

NEW PERSONAL DATA

One new type of personal data comes from spit. Each person's saliva contains

Some companies collect saliva or other samples from people to provide information about their ancestry.

deoxyribonucleic acid (DNA). DNA is a chemical found in nearly every cell in the body. It carries genetic information. Genetic information determines some of a person's traits, such as eye color. Some DNA is passed down from parents to their children.

Some companies collect people's DNA. People mail saliva samples to these companies. The companies study the DNA in the samples. DNA can give information about which parts of the world a person's ancestors are from. It can also give information about certain diseases or health conditions a person might be at risk of developing. The companies save this data. They sell it to drug companies. Drug companies can use it for medical research.

People may find more uses for DNA data in the future. Scientists could use it to develop cures for diseases. But some people have concerns. They think this data could be used to discriminate against people. For example, the Canadian government uses DNA data to investigate immigrants. Some people also think DNA companies have confusing data-sharing policies. Customers might not know that their data can be sold or stored for years.

DATA LAWS

Some people are trying to set new data protection standards. In 2018 the European Union (EU) passed the General Data Protection Regulation (GDPR). This law requires companies to tell people about the data they collect. It allows customers to ask for their data to be deleted. It requires companies to tell customers about data breaches.

PERSPECTIVES
DNA DATA

Subodh Bharati is a Canadian human rights lawyer. He found that the Canadian government was collecting personal data from DNA ancestry companies. The government used this data as evidence to prove that some people were not citizens. The government deported these people, or removed them from the country. Bharati believes that using DNA data to determine someone's citizenship is unfair. He said, "It's clear DNA doesn't give someone's nationality."

Vera Jourova, the EU Commissioner for Legal and Consumer Protection Policy, gave a press conference about the GDPR in 2018.

The GDPR also addresses unfair data collection

and usage. It tries to prevent data-based discrimination.

For example, Google collects personal data. It uses

this data to create targeted ads. Researchers found

that Google showed different job ads to men and

women. Men received more ads about high-paying

Today, some people continue to protest for better data privacy laws.

jobs than women did. The GDPR allows customers to

protest discrimination in data collection and usage.

Customers can ask how companies are using their data.

The GDPR has effects beyond the EU. Many data-collecting companies operate globally. Data standards worldwide may change as standards in Europe change. In the United States, there is already a push for similar laws. In 2018 California passed the California Consumer Privacy Act. This law is similar to the GDPR. It gives consumers the right to ask for their data to be deleted. Organizations that break data privacy rules are forced to pay a fine.

Personal data collection is a powerful source of information. It can be used to do research that benefits people. It can be used to

PREDICTIVE POLICING

Police can use personal data to identify people who may commit crimes. This process is called predictive policing. Police may target people who have arrest records. Many people worry that predictive policing could lead to racial discrimination. African Americans are more likely to have arrest records than people of other races. For this reason, police may target them.

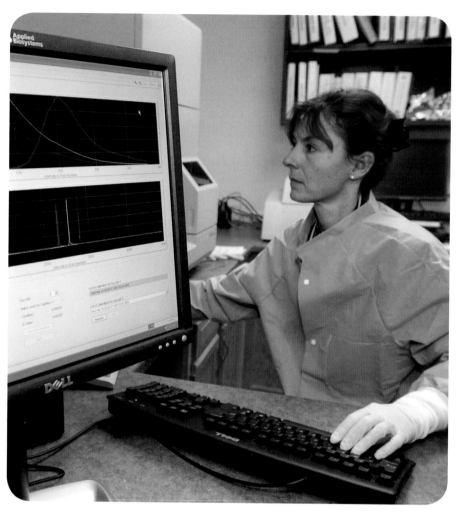

DNA databases can help identify people who have committed crimes. Scientists try to match DNA samples to records in the database.

make predictions. But data can be collected unfairly. People today are trying to develop ethical ways to collect and use personal data.

STRAIGHT TO THE
SOURCE

Teresa Scassa is a law professor at the University of Ottawa. She studies data protection. In a 2018 interview, she explained common data privacy issues. She said:

> *Some level of data sharing is often necessary for most goods and services we purchase or license. The more we want services customized to our interests, the more data we may need to share. Data is also increasingly the informal method of payment for many "free" digital services. Consumers may not be aware just how much data, what kind of data and how much detail they are providing. While there are contexts where individuals voluntarily share personal information—such as over social media—there are many other contexts in which the sharing is difficult to characterize as truly voluntary.*

> Source: Tim Sandle. "Q&A: Data Ownership Conundrum in the Data Driven World." *Digital Journal*. Digital Journal, September 25, 2018. Web. Accessed December 19, 2018.

Back It Up

The author of this passage used evidence to support a point. Write a paragraph describing the point she was making. Then write down two or three pieces of evidence she used to make this point.

FAST FACTS

- Personal data is information about people that can be collected and stored. Names, email addresses, online shopping histories, and medical records are all types of personal data.

- Personal data collection has expanded in the past few years. People are creating more data and collecting it in new ways.

- Some researchers study personal data to make predictions. These predictions can benefit people.

- Some companies sell their customers' personal data to make money. They use the data to learn how to better market their products to customers.

- Some hackers try to break into company databases and steal personal data.

- Some people think personal data collection violates their right to privacy. Others worry that personal data could be collected unfairly or used to discriminate against people.

- As technology develops, new types of personal data and ways of collecting it are being created.

- Activists are working to pass new data laws. The purpose of these laws is to protect user privacy and ensure that data is used fairly.

STOP AND
THINK

Tell the Tale

Chapter Two of this book talks about data breaches. Imagine you were the victim of a data breach. Write 200 words about your experience. How do you think the breach could have been prevented?

Surprise Me

Chapter Four discusses data selling and sharing. After reading this book, what two or three facts about data sales surprised you? Write a few sentences about each fact. Why did you find each fact surprising?

Dig Deeper

After reading this book, what questions do you still have about personal data collection? With an adult's help, find a few reliable sources that can help you answer your questions. Write a paragraph about what you learned.

Welcome to you®

HEALTH + ANCESTRY

saliva collection kit

Take a Stand

Some people think that personal data is a valuable research tool. Others think data collection threatens their privacy. What is your opinion? Do you think there should be limits to data collection?

GLOSSARY

ancestry
a person's relatives from earlier generations or ethnic descent

anonymous
not named or identified

app
short for application, a computer program that has a special function

bias
favoring certain people or points of view over others

consumer
someone who uses a product

data
information that can be stored and studied

discrimination
the unjust treatment of a person or group based on race or other perceived differences

hacker
someone who breaks into secure systems or databases

immigrant
a person who moves from one country to live in another

terrorist
someone who uses violence or threats to frighten people

ONLINE
RESOURCES

To learn more about personal data collection, visit our free resource websites below.

Visit **abdocorelibrary.com** or scan this QR code for free Common Core resources for teachers and students, including vetted activities, multimedia, and booklinks, for deeper subject comprehension.

Visit **abdobooklinks.com** or scan this QR code for free additional online weblinks for further learning. These links are routinely monitored and updated to provide the most current information available.

LEARN
MORE

Gifford, Clive. *The Science of Computers*. New York: Crabtree Publishing, 2015.

Smibert, Angie. *Inside Computers*. Minneapolis, MN: Abdo Publishing, 2019.

INDEX

About the Author

A.W. Buckey is a writer living in Brooklyn, New York.